Nature of Vedic Science and Technology

By

Dr. Ravi Prakash Arya

INDIAN FOUNDATION FOR VEDIC SCIENCE

H.O.1051, Sector-1, Rohtak, Haryana, India Ph. 01262-292580
Delhi Contact Ph. Nos.: 011-65188114; 09313033917
Emails: vedicscience@rediffmail.com
vedicscience@hotmail.com
Website : www.vedascience.com

First Edition

Kali era: 5015 (c. 2014)

Kalpa era: 1,97,29,49,115

Brahma era: 15,50,21,97,9,49,115

ISBN No. 8187710632

Table of Contents

1

Introduction to Vedic Science and Technology

Vedic literature is the most significant contribution of India to the world and through it, India has contributed emensely to the various fields of arts, sciences, technology and spirituality. It is here in this country that a systematic attempt has been made to understand the human speech thousands of years ago while the rest of the world was oblivious of communicaton styles and skills. Remarkable contribution of this country can be counted towards the origin of human speech, grammar, numerals, mathematical sciences, astronomy, arts, archaeology, sculpture, social sciences, bio-sciences, positive sciences, high level of technology and spiritual sciences. A country, which has such a sound mathematical base, has naturally to pursue science and technology for improving the quality of life of her people.

William H. Gilbert in his book "People of India" writes very clearly-

> "In history of human culture the contribution of the Indian people in all the fields has been of the greatest importance. From India we derived domestic poultry, shellac, lemons, cotton, pepper, jute, rice, sugar, indigo, buffalo, cinnamon, ginger, sugarcane, the games of chess, pachisi, polo, the zero concept, the decimal system, the basics of certain philological concepts, a wealth of fables with moral import, an astonishing variety of artistic products, and innumerable ideas in philosophy and religion such as asceticism and monasticism."

Will Durant, an eminent philosopher of USA observes-

"India is the motherland of our race and Sanskrit the mother of our European Languages. She was the mother of our philosophy, mother, through Arabs, of much of our mathematics, mother through Buddha, of the ideal embodied in Christianity, mother through the village communities of self government and democracy. Mother India is in many ways Mother of us all."

The Vedas have been the fountainhead of all types of knowledge, science and technology. To understand the development of science and technology in Vedic age, it is imperative to understand the Vedas and Vedic science vis a vis modern science.

Vedas and Vedic Science

When we talk about Vedic Science, It is but natural to have a quest about the Veda.

When we talk about the Veda, we, generally, have a picture of a holy scripture in mind. We think that Vedas are of the same category of literature as that of Bible or Koran. People, generally, think that Vedas are related to Hindus as the Bible and Koran to Muslims and Christians. Just as the Bible and Koran are religious scriptures used for performing the various rites, ceremonies concerned with the compartmentalized life of persons of particular communities, Vedas are also the religious scriptures used for performing various ceremonies and rites of Hindus. In fact, this view received its first sanction at the hands of various western scholars who started the study of the Vedas having regarded them as religious scriptures. This is evident from the Max Müller's work called as 'The Sacred Books of the East Series'. This is how the Vedas are also categorized in the list of other sectarian or communal literature. Vedas are, in fact, the document of the great advance made first ever by humanity on this globe when there originated no sect or religion on the earth. The Vedas are the results of Vedic seers' comprehension of the spiritual, biological, physical and inter-stellar world. Vedas are regarded sacred, not because that they are related to a particular community or because they were composed by a prophet of a particular community, but because they are the store house of all the true knowledge, because they represented the science of the entire cosmic life tested, realized and visualized by the great seers of India; because they contained universal laws irrespective of cast, creed, race, religion or region.

In the same line, it can be stated that Vedic science is the science of creation, representing all its three aspects, i.e. *adhibhautika* (physical or worldly), *adhidaivika* (astronomical or astrophysical) and *adhyātmika* (metaphysical or spiritual). Thus in more vivid and lucid terms, it can be stated that Vedic science is a science:

1. That defines the relationship between mind, body and soul at the *adhyātmika* or metaphysical level.
2. That defines the phenomena of life, death and hereafter.
3. That defines the relationship between the biological life and the physical life in the context of earth and cosmos beyond.
4. That defines the relationship between the living body and the cosmic body.
5. That defines the parallelism between physical, astrophisical and metaphysical levels of creation. In other words, it defines the relationship between physics, astrophysics and metaphysics.

To sum up, one can say that Vedic Science is a science of earth, a science of cosmos and a science of spirituality or consciousness.

Now-a-days there is a great demand from all quarters that methodology and working of Vedic science may be defined in terms of and relation to modern advancement in science and technology, simply because everyone of us is very well acquainted with the methodology, laws and working of modern science. Comprehension of Vedic science will be faster and quicker if it is defined vis-a-vis the modern science. In view of the same fact, various aspects of Vedic Science are defined hereunder in relation to modern science.

2

Vedic Science
vis-a-vis
Modern Science

Vedic Science is a science based on realization. Modern science is based on observations. Whatever is observed during experiments in a laboratory is often incorporated in the parameters of science. As far as Vedic Science is concerned, it is not merely based on observations; rather based on realization or visualisation in addition to experiments. Realisation has far deeper meaning than the experiment. Comprehension of Spiritual aspect of the creation is possible through realisation. As such, whatever is realized or visualised becomes science in addition to what is observed. Defining the science, Ādi Saṅkarācārya says: *jñānaṁ viṣaya viṣayānubhūtir vijñānam.* That is jñāna is bare information or knowledge of an object or a thing, but science is the realization of that particular thing or an object. Maharishi Dayananda Saraswati, one of the leading Vedic scholars and social reformer of 19th century also defines Vijñāna as: *tasya parameśvarādārabhya tṛṇaparyanta padārtheṣu sākṣād-bodhānvayatvāt,* i.e. domain of science entails the realisation of all the things right from the root of grass till God. Thus, we see that realization is a deeper concept than observation. Observational power is limited. Everything cannot be observed, e.g. observation of abstract things is not possible. On the contrary, realization of even those things which cannot be observed is possible. Observation is a physical concept whereas realization is a spiritual concept. Realization is oneness with the thing or an object realized. When a Yogī realizes Brahman, he becomes virtually a Brahman - *yo Brahma jānāti sa brhmaiva bhavati.*

A Perfect Science: Vedic science is a perfect science as compared to modern science. The Vedic scientific advancement, which was branded by the modern scientists as mere fiction based upon conjectures, has now been acknowledged as truth. The more the modern science advances, the more validity of Vedic science is established. Here are some facts, which may distinguish modern science from Vedic Science.

Modern science has no perfect time reckoning system, whereas the Vedic science has a fully developed time reckoning system based on astronomy. We do find modern geologists talking about some geological eras spanning around 250 crore years, but we find in the Vedic age the biggest ever era spanning into 311,040,000,000,000 (311 Trillion) years representing the total age of the universe from its origin to end. Modern science has done some exploration at the physical and astronomical level, but it has no knowledge about metaphysical creation. It has not been able to define the relationship between the mind, soul and body. It has no technology so far developed to prolong the age of human beings. The life and death is still a mystery in the modern science. There is no solution of the problem of hereafter in the modern science. In modern science, nothing is said with certainty, everyday the theories are changed or updated. Whereas, the Vedic scholiasts talk about with the gist of certainty.

Direct knowledge: Modern scientist conducts experiments in laboratories and what is observed in the laboratory experiments is taken as science. Thus the laws of modern science are framed on the basis of indirect knowledge gained through the laboratory experiments. On the other hand the Vedic scientist had the highly developed intelligence capable of gaining the natural laws directly. For him the whole universe itself was a laboratory. He was capable of perceiving what was happening in the nature directly without any aid or external help.

Exact dating: The conclusions of modern science are based on probability. Modern science doesn't have exact dating regarding the age of universe, etc. It presents estimates in round about figures. However, Vedic seers have given exact figures whether it is the date of the origin of universe or galaxies or solar system or earth or moon or what not. As far as modern science is concerned, it has no hesitation to forego the difference of 20 to 50 Million years if the time spans over to 200 Million years. Similarly, if there is the question of 60 to 100 Million years, the modern science readily accepts the exemption of 14 to 15 Million years. In case the calculation extends to 500 Million years or more, the exemption of 100 Million years is permissible. As far as the Vedic science is concerned, no such exemption is permissible whether there is calculation of 20 Million years or 2 Billion years.

Inductive Science: Modern science adopts the deductive method of inquiry, i.e. it goes from part to the whole, whereas Vedic science adopts the inductive method of inquiry. Rather going from part to whole, it tries to comprehend the whole. If the whole is comprehensible, the parts will be comprehended automatically. Nevertheless, it would be very difficult to comprehend the whole based on the parts. It would completely be a conjectural work to determine the shape and structure of a complete body based on parts without having the prior knowledge of complete body. It will be more or less like the story of four blinds set out to determine the shape of an elephant. The blind who caught hold of the elephant by teeth, described the elephant like the sword. The one who caught hold of it by tail described it as the broom. Who caught hold of the elephant by legs, described it as a pillar. The one, who caught hold of the elephant by trunk, described it as a pipe. Thus, they made a puzzle of elephant. A person who had perfect vision solved their puzzle. He told them that elephant is like a sword, a pipe, a pillar as well as like that of a broom. All these make a complete elephant.

Science of whole: As indicated earlier, the Vedic science is the science of whole or Pūrṇa. The comprehension of the whole takes place faster, quicker in a more perfect and convincing manner. For instance, there will be a total difference of perceptions of the earth, when it is perceived as a whole from the space and when it is perceived from the earth itself in parts. The perceiver of the earth from the space will have a more accurate, precise and quick comprehension regarding the size or appearance of the earth, than the one who tries to construe it from fragments. He may take years and even then, he may fail to draw the right conclusions. For the inductive method of inquiry, one has to develop his senses to the extent that he is capable to perceive the whole. If one wants to comprehend the universal system through the deductive methods, he may have to take innumerable births to reach the ultimate reality, but the inductive method will make one to comprehend the whole system within one day. For example if you develop the habit of reading the page as a whole, your reading will be faster and quicker than the one who reads it word-wise and line-wise.

Upgradation of human level of understanding: Vedic science is the combination of practices to upgrade the human level of understanding and thinking to a level at which the universe functions itself and once a human being reaches that level of understanding, he finds the functioning of universe as simple as watching a documentary film. Vedic science deals with different disciplines, which are interlinked. For example, medicine is linked with astronomy, astrology, psychology, music, aroma therapy, generics and food habits. Health of individual or family depends largely on social stability in terms of commerce and military. Thus, macro level goals can be achieved by success at micro level.

On the other hand, modern science is the skill to formulate everything in a simple way that human can understand. We can simulate this process as breaking complex task into different parts that can be easily understandable. So modern science allows a human to be expert in some part say chemistry, some will be expert in other part say physics as some on mathematics. If we look at the modern science, everything is specialised and specialists are able to answer most of the puzzles in their respective fields only. They rely on senses and instruments, which provide a minute atomic view of the entire knowledge.

3

Nature of Vedic Science and Technology

As the Vedas are the oldest literary records of great advance made by humanity in the world. The sublime and the rich poetry of the Vedas, the antiquity, perfectness, copiousness, exquisite refinedness and wonderfulness of the structure of the language they are composed in all point out to the highly cultured, and civilized nature of the race within which this rich, and everlasting heritage was originated. The term civilization is itself a pointer to an advance stage of living, where the people, are able to lead a comfortably settled and prosperous life and can spare time for innovative activities in various fields.

Under the circumstances, it is but natural for modern generations to evince interest for the social conditions of the highly advanced race that is said to have represented the remotest past of this country. Now, when the science and technology have become the way of life and everything is tested and examined in the light of scientific ideas, it becomes more interesting to know about the nature and type of the science the Vedic people professed and the nature and type of the technology they developed and devised.

Origin of science: The origin of science always takes place in the outlooks of individuals or the society towards the creation of the world of living (*Cetana*) or non-living (*Jaḍa*) things. The innovations, inventions or discoveries in the field of science and technology take their shape or derive inspiration mainly from the philosophical hypothesis formulated by an individual or the society towards the origin, or composition of the world. It depends entirely upon the philosophical viewpoint of the innovator, inventor or discoverer as to how to philosophize the world around him, whether he takes it as a composition of *Prakṛti* (matter) only or as an outcome of a mere accident of *Puruṣa* (self), and *Prakṛti* or as an outcome of accident of Puruṣa and *Prakṛti* caused by the will of Almighty *Brahman.* So to say, one's range of innovations, inventions or discoveries is, always encompassed by the range of one's philosophical hypothesis. For instance, in the modern times, the living, or non-living world is considered to be the creation / composition of matter. This is why, the modern science is studied in the name of' matter, physical properties of which are studied in Physics and Chemical properties in Chemistry. Modern technology is also devised based on the advancement made by the material science or positive sciences, viz. Physics and Chemistry.

The Vedic Origin of Science: In the Vedic times also, science was developed around the philosophical innovations of the Vedic *Ṛṣis.* According to philosophical concept of the Vedic *Ṛṣis*, the world was not the composition of matter only, it was considered to have represented by *Prakṛti, Puruṣa* and *Brahman* in various capacities.[1] It evolved on account of the accident of Puruṣa and *Prakṛti*, the accident being materialized by the will (*Saṁkalpa*) of the *Brahman.*[2] The accident of *Puruṣa* with *Prakṛti* which is the name of the equilibrium of *Satva, Rajas* and *Tamas* qualities, brings about the disequilibrium in triad of qualities (i.e. *guṇatraya*) which ultimately causes the *Vikṛti* of *Prakṛti*, or evolution of matter as under :

> *Prakṛtermahān mahato 'haṁkāro 'haṁkārāt pañcatanmātrāṇi ubhayamindriyaṁ pañcatanmātrebhyaḥ sthulabhūtāni puruṣa iti pañcviṁśatirgaṇaḥ*

The process of the evolution may be illustrated as under:

Will of Brahman

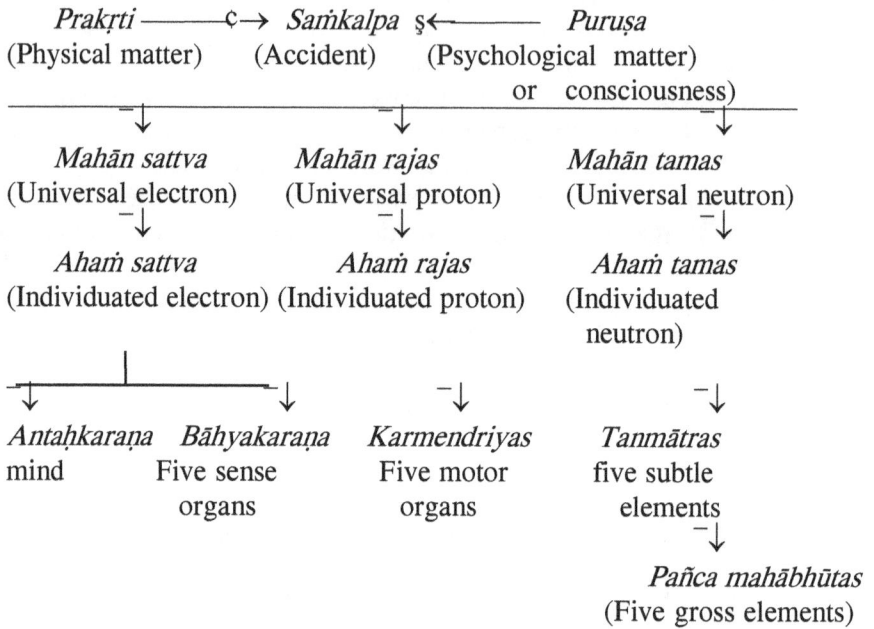

Prakṛti ———ç→ *Saṁkalpa* ş←——— *Puruṣa*
(Physical matter) (Accident) (Psychological matter) or consciousness)

↓ ↓ ↓

Mahān sattva (Universal electron) | *Mahān rajas* (Universal proton) | *Mahān tamas* (Universal neutron)

↓ ↓ ↓

Ahaṁ sattva (Individuated electron) | *Ahaṁ rajas* (Individuated proton) | *Ahaṁ tamas* (Individuated neutron)

↓ ↓ ↓ ↓

Antaḥkaraṇa mind | *Bāhyakaraṇa* Five sense organs | *Karmendriyas* Five motor organs | *Tanmātras* five subtle elements

↓

Pañca mahābhūtas (Five gross elements)

Thus based on the foregoing discussion, it can be summed up that three types of elements play their various vital roles in the creation of universe. For instance:

(1) The *Prakṛti tattva* which evolves into *tanmātras* (pure *bhūtas*) and *sthula bhūtas* (gross or standard elements) is the *upādāna* (material) cause of the creation of living and non-living world, as the whole world is created as an outcome of the evolution of *Prakṛti*. Actually, *Prakṛti tattva* is the same althrough, whether It is a living or non-living world. Perhaps due to the same reason it was theorized by ancient *Ṛṣis* (researchers): *yad aṇḍe tad brahmāṇḍe*. 'Whatever is contained in the physiology is embodied in the physical world outside.[4] In other words, the *Prakṛti* within the living beings is closely attached with the without non-living world.

(11) Ātma *tattva* or *Puruṣa viśeṣa* (Individuated soul) is the simple cause of the creation.

(111) *Puruṣa tattva* or Brahma (universal soul) is known as the

efficient cause of the 'whole creation. Its accident with *Prakṛti* causes the evolution of universe.

Above-mentioned Vedic philosophical concept of the creation of world points out to three possible stages of evolution of philosophical / scientific viewpoint towards the world.

At the first stage which may be called the stage of sensory perception, a philosopher conceives the world as merely a *prākṛtika* / material evolution due to ignorance (*ajñānatā*). This stage is known as the stage of' utter ignorance. At this stage his attempt of doing science will be the perfection of physical and chemical properties of the *Prakṛti* / matter and so will be evolved the material science known as positive sciences, viz. Physics and Chemistry. And on the basis of the positive sciences, the type and nature of technology developed will be the one called as *Yāntrika*, that which pertains to *Yantras* / machines developed out of matter. At the second stage, apart from the stage of utter ignorance, one is able to come out of ignorance and can understand the role of *Ātma tattva* and *Brahman* in the creation of the world. One is also able to recognize the presence of *Brahman* as an indifferent witness to the going about of the created world and recognizes the proximity of' *Ātma tattva* with that of the *Prakṛti tattva* within himself. At this stage, instead of doing with the outside *Prakṛti*, he tries to perfect the self-nature or the *Prakṛti* within.

This process is known as *ātma-siddhi*, or self-perfection, or self-accomplishment, which can be achieved through *Yoga*, or say through regulation and concentration of mind (*samādhi*). The *Yoga,* which is also known as *Ātma Yajña* helps a Yogī in perfecting the *antaḥ-prakṛti* or nature within / bodily nature. Actually there are two stages of *samādhī*, *saṁprajñāta* and *asaṁprajñāta*. At the *saṁprajñāta* level of' *samādhī*, a Yogī maintains his separate identification or individuated existence and is not able to become one with the Almighty.

During this phase of his practice, a Yogī is able to, attain self-perfection or self-accomplishment, and this attainment, as it has already been said that the *Prakṛti tattva* is the same althrough regardless the distinction of body or universe, accomplishes him to have a perfection over the *Prakṛti* without. At this stage he develops extrasensory perception and is endowed with several divine powers on account of various perfection achieved through regulated and concentrated mind[5] e.g. divine vision, divine audition, divine olfaction, divine skin sensation, divine energy, divine motion and speed, etc.[6] So at this stage, his sphere of doing science won't be the material one. The concepts of light, heat, energy, sound, speed, motion, force, electricity, attraction, and gravitation, etc. which are dealt with under material sciences, or positive sciences will change into the concepts to be dealt with under Bio-science. Now, he will deal with such concepts as bio-light, bio-heat, bio-energy, bio-sound, bio-speed, bio-mass, bio-force, bio-electricity, etc. Based on these concepts of Bio-scicnce, he will (develop) a technology altogether different from the one known to a nonprofessional. That will often appear to a layman as miraculous, or magical one. For instance, television will find a replacement in the divine vision; telephone or telegraph Will be replaced by telepathy; different kind of transport, such as air, marine and surface transports and road and rail technology will have their substitutes in the divine powers of *utkrānti*, *jalāsaṅga*, *paṅkāsaṅga* and *kaṇṭakāsaṅga*, etc.[7] as mentioned in the *Yogadarśana.*

Thus, the technology at this stage takes the form which is known as the *Tāntrika* one.

The third and final stage is that of *asaṁprajñāta samādhī* in which a *Yogī* completely identifies himself with the God and a sense of duality is lost. It may be known as a stage of self-perception. It is a state of complete bliss, a state of complete union with God. It is a state described in *Upaniṣads* as:

Ahaṁ brahmāsmi 'I am *Brahma*',

Tattvamasi 'Thou art He';

Sarvam khalu idam Brahma neha nānā asti kiñcit 'All that is visible is Brahma, nothing is here of different kind.'

At this stage of non-duality, when a *Yogī* becomes virtually Brahman, he develops all such qualities as are thought to be the Godly ones. Thus the concepts of material science / positive science that developed into the concepts of' Bio-science at the second stage, culminate at the third stage into the absolute ones, such as absolute light, absolute heat, absolute energy and absolute mass, etc. He, then, becomes able to do or materialize anything instantly at his will like that of the God himself. In other words, he deals with the science of the absolute ones. His technology finds an unending and infinite source in the form of absolutes. Actually, this is the source which will remain absolutely unaffected how so much it is exploited. This fact finds a nice expression in the utterance of an Upaniṣadic Ṛṣi as :

pūrṇamadaḥ pūrṇamidam pūrṇāt purṇamuducyate

pūrṇasya pūrṇamādāya pūrṇamevāvśiṣyte.

'That is, if absolute is taken out of absolute, the remaining will be the absolute one.'

The significance of this concept of absolutism has also been recognized by the modern science, hence the same has been accepted as infinity. But it finds its application (∞∞ - ∞∞ = ∞∞) merely in theory rather than in practice. Hope the further advancement in modern science will be able to discover the Bio-scientific values of the concepts that are dealt with in the positive sciences and will also be able to recognize the importance of infinity at a wider and larger scale of its applicability. The nature of this type of science and technology is described as *Māntrika* which is more often then not enshrined in the four corners of *adhyātma*.

All these three types of science and technology had found their recognition and codification in the whole Vedic literature. For instance, all the Vedic mantras signify to *adhyātma*, *adhidaivata* and *adhibhūta* which further support the concept of *Mantra*, *Tantra* and Yantra respectively. The composition of the later Vedic literature, e.g. *Brāhmaṇas*, *Āraṇyakas*, *Upaniṣads* and *Sūtras* was carried out keeping in view the above concept.

Six systems of Indian Philosophy also take up for consideration the one or two subjects from the above mentioned three concepts of science, e.g. *Vedānta* or *Uttar Mimāṁsā* deals exclusively with the *adhyātma*; *Vaiṣ eśika* of' Kaṇāda is an elaboration in positive sciences (*adhibhūta*);

Sāṁkhya and *Yoga* cover an *adhidaivata* aspect; *Nyāya* takes up the methodology of doing science and *Pūrva Mimaṁsā* deals with the *Yāntrika* mechanism of ancient science. Similarly, three types of Vedic technology, find their mention in the books dealing with technology. For instance *Bṛhadvimānaśāstra*, an ancient Vedic treatise on the science and technology of aeronautics, composed by Maharṣi Bhāradvāja makes a mention of three types of *Vimānas* (airplanes) as *Māntrika*, *Tāntrika* and *Yāntrika* ones devised on the basis of the science of *Mantra* (*adhyātma*), *Tantra* (*adhidaivata*) and *Yantra* (*adhbhūta*) respectively.

The Development of Science and Technology in the Vedic Age

Although all the three types of science and technology were practised by the Vedic people althrough their age, yet in the earliest phases of Vedic life, the science of *Mantra* and the *Māntrika* technology was all the more practised by the high spirited Vedic Ṛṣis. Actually, the term Ṛṣi itself signifies to the researcher who is dealing with the science of *Mantra*. *Ṛṣ ayo mantradṛṣṭāraḥ* or *Ṛṣirdarśanāt*. 'Ṛṣis were those who had the perception of *Mantra*. *Mantra* or *Śruti* being the direct or primary knowledge acquired by the Ṛṣi or seer during the course of self-perception. *Mantra* being related with the science of God, was also known as *Brahma*. The concept of *Śabda Brahma* in the philosophy of grammar takes its shape from the above quoted Vedic innovation of *Mantra-Brahma*. The Ṛṣi who knew or did the science of *Mantra* or *Brahma* were known as *Brāhmaṇas*. *Brahma jānāti Brāhmaṇa*. The technique of devising or developing the *Māntrika* technology based on the science of God/*Mantra*/*Brahma* was called the *Brahma-Yajña*. The above mentioned science of *Mantra*/ *Brahma*/ God and the technique of *Brahma Yajña* was incorporated into the frame work of *adhyātma*. All the Vedic *Mantras*, as per traditional convention, primarily signify to *adhyātma*, *adhidaivata* and *adhibhūta* being the secondary significance. The *Brāhmaṇas*, which are the explanatory notes on the Vedas, refer to the sense of *adhyātma* 100 times, whereas *adhidaivata* and *adhibhūta* have been referred to 60 and 9 times respectively. Actually, *adhyātma* was the monopoly of a few high spirited Ṛṣis and its practice by layman was not possible. So, in the ancient times, the common man remained aloof, unfamiliar or unacquainted with the highly advanced and sophisticated nature of this science. In fact, it was a subjective concept which could not benefit those subjects who were not able to practise it. This is why, as is gathered from the stories of Epics and *Purāṇas*, the people who were keen to learn this science, had to do penance to please such Ṛṣis as were well equipped with this science in order to learn it from them. In fact, the word *Smṛti* is suggestive of the indirect or secondary knowledge that was made to realize by Brahmaṛṣis, having *Śruti* knowledge, to later Ṛṣis.

On the contrary, in the later phases of Vedic life, owing to the decline in the high values and concepts of spirituality, the practice of non-duality and oneness with God ceased to be observed and thus went out of vogue. This gave rise to the decline of the tradition of Brahmaṛṣis. The sorry decline of this noble tradition has also been alluded to by Yāska as :

sākṣātkṛdharmāṇaḥ ṛṣayo babhūvuḥ. te avarebhayḥ asākṣ ātkṛtdharmabhyaḥ upadeśena mantrān samprāduḥ [9]

> *"That is, there were born Ṛṣis to whom was revealed the science of* Mantra *(*Dharma*). They preached this science by oral, instruction to their juniors who were devoid of this type of revelation.'*

With the degradation in the high values of spirituality and the decline of the noble tradition of *Brahamrṣis*, there arose a new generation of *Ṛṣis* whose concept of spirituality remained confined up to the observance of *samādhī* to its *asaṁprajñāta* level, i.e. the level at which they were endowed with divine powers or extra-sensory perception. On account of having endowed with only divinity, they were often called as *Devas*. We come across numerous references in the *Brāhmaṇas* and *Āraṇyakas* where the *Devas* are differentiated from *Manuṣyas* on account of their being *ūrdhvaretas*. Thus the later Vedic epoch records the currency of the science of Soul or *Tantra* achieved through the technique of *Ātma-Yajña*. The science and technology based on *Tantra* being divine (*daivika*) in nature and appearance, falls in the sphere of *ādhidaivika* activities. But we find a further decline in the concept of spirituality, the people having been confined to the divine aspect (or *siddhi*) of the *Yoga*. Actually, the attainment of *siddhi* or divine power of some sort or other is not very difficult to achieve through a little practice of *Yoga*. In fact, these *siddhis* or divine powers are described by Patañjali, the composer of *Yoga Darśana* as the hurdles on the way of real progress of the Yogī, as the Yogī often endowed with these powers gets astrayed from his actual path and is led by the mortal or worldly longings for name and fame. At this stage a Yogī is described or christened as *Bhraṣṭa* Yogī or the fallen one.

This concept of *Bhraṣṭa Yoga* presents a distorted picture of the science and technology of *Tantra* in the post Vedic period. The people are often seen to have indulged in the demonstration and performance of *māyāvi śaktis* or magical powers achieved sometime through *Yoga* and sometime through the knowledge of specific chemical or physical properties of the material things which were unknown to both the learned or laity in the society. Numerous references as to this regard of practices can be gathered from the Epics and *Paurāṇika* literature.

The increasing practice of the black rituals like that of *Māraṇa* (to kill someone), *ucchāṭana* (to trouble Someone), *sammohan* (to render someone unconscious) and *vaśikaraṇa* (to tame someone for his use) in the public life also exhibits decline in the standard and distortion of the healthy tradition of science of *Tantra*. In fact, this distorted form of Tāntrika tradition is represented still by the present day magicians and miracle-men. The science of *Tantra* though could find a wider sphere as compared to the science of *Mantra*, could also not percolate to the common man in society. So, the common man still remained aloof of the actual significance of the science of *Tantra* and thus deprived from the benefits which could be availed from the *Tāntrika* techniques. Moreover, he became the victim of clever tricks of black magic often played by the so-called *Tāntrikas*. Actually, this type of ritualism was responsible for the decline of Āryanism or Brāhmaṇisam and for the origin of Buddhism and Jainism in this country.

Alongside the currency of the science and technology of *Mantra* and *Tantra*, a need for evolving the technology which could save the common man from natural calamities was eagerly felt. So the science of nature / atmosphere was also evolved by the Vedic people. It found its genesis the technique of *Deva-Yajña* form of *Yajña*. *Deva-Yajña* was an attempt at perfecting the elements of nature in order to bring the nature down to suit the requirements of living-beings on the Earth. Its main objective was the *devapūjā*, i.e. to please the mid-spatial and celestial elements, so that they may bestow their favour upon the living-beings and thus give rise to the process of *Agnihotra* or *Haviryajña*.[8] The *Yajña* was also aimed at *saṁgatikaraṇa*, i.e. to create something new out of the material (physical and chemical) composition of an element or elements. The material requirements also compelled the Vedic people to do the science of matter and develop and devise the *Yāntrika* technology which was also known as *Bhūtayajña*.

Actually, in the post Vedic period when the science of *Mantra* almost went out of vogue due to its subtleties and the science of *Tantra* also began to fizzle out owing to the distortion of its pure and original nature, it was only the *Yāntrika* science and technology that could eventually make its headway, perhaps, on account of its easily understandable nature. At the initial stages though for the namesake it assumed a separate and distinct form as a science, in actual practice it remained combined with the *Tantra*. That is, it was practised under the influence of *Tantra* and so the *Yāntrika* technology was also operated with the help of some *Mantra* (i. e. divine sound), etc. We come across such references as to show full development of this science in the Epics, which make a mention of sophisticated type of weapons that were used in the battlefields and were operated by some divine sounds (*Mantras*). This was known as the *abhimantrita* use of *Yantra* (machine / weapon). But with the passage of time *Yāntrika* science and technology assume its pure form and no longer depended upon *Tantra* for its operation. The extant *Bṛhadvimānaśāstra* and the *Yantrasarvasva*, quoted by the *Bṛhadvimānaśāstra*, are living testimony to the highly advanced nature of ancient Indian *Yāntrika* science and technology. Both of these treatises give a detailed break-up of scientific advancement and technological know-how in the field of aeronautics, machines and other scientific instrumentation.

The *Bṛhadvimānaśāstra* deals with the theory and practice of three types of *Vimānas*. *Māntrika*, *Tāntrika* and *Yāntrika* ones which could be operated upon land, in waters or in the sky as needed. The book, on the basis of *Yantrasarvasva*, talks about the installation of various types of instruments in order to save the aeroplane from various dangers. For instance, *Śiraḥkilaka yantra* is described to dispel the effect of lightening. *Varṣopsaṁhāra*, *Trāyasyavātanirsana* and *Ātapopasaṁhāra* instruments to neutralize the effect of rain, winds and heat respectively on the airplane. In fact, the extant *Bṛhadvimānaśāstra* is a book that provides a valid answer to the magical warfare often described to have been fought in ancient times. Moreover the same book records the name of some forty personalities who were engaged in the science and technology of aeroplanes. Further breakthrough in the advancement of *Yāntrika* science and technology of ancient India is possible if and only if such an extant literature on ancient sciences as *Bṛhadvimānaśāstra* and *Yantrasarvasva* is thoroughly and practically studied and other such literature as has been lost in the hoary past with the passage of time is searched out.

Notes and References

1. See *RV.* 1.26.4.

   dvā suparṇā sayujā sakhāyā samanaṁ vṛkṣaṁ pariṣasvajāte
   tayoranyaḥ pippalaṁ svādvattyanaśnanyo, abhicākśiti.

   'I.e. two beautiful friendly birds (*Ātman* and *Brahman*) are sitting together on a tree (*Prakṛti*). One of them tastes the sweet fruits and another sits indifferently looking at or acts as a mere witness to the former'.

   See also *Chāndogyopaniṣad* :

   asadvā idamagrāsīt, ātmā vā idamagrāsīd, brahma vā idamagra āsīt

   'i.e. before the creation, *Prakṛti* existed. *Ātmā (Puruṣa)* existed and *Brahman* exited.'

2. Cf. following statements of *Upanisads* : *eko'haṁ bahusyām* 'I am alone, let me manifest in many.'

   tadaikṣat bahusyām prjāyeyeti. (He wished to be many, so gave birth to *prajā*'. *So'kāmayat bahu syām prajāyeyeti* : 'He wished to be many, so gave birth to people.'

   See also Gitā :

Saṁkalpaprabhvān kāmān sarvān viddhi. 'All things should be known to have been the creation of Will.'

3. *Sattvarajastamasāṁ sāmyāvasthā prakṛti,* (*Sāṁkhya Darśana* 1.1)
4. *vide supra* 1.2.1
5. *yuktena manasā vayam devasya savituḥ save svargyāya śaktayā.* (*VS*.2.2)
6. Cf. for details Patañjali's *Yoga Darśanam* .
7. *udān-jayājjala-paṅkakaṇṭkādiṣvāsaṅga utkrāntiśca. Yoga Darśana* 3.39
8. Cf. for detailed break-up of this form of *Yajña*, see '*Vedic Meteorology*' by the present author, published by Parimal Publications, Delhi, 1995.

4

Role of Science : Vedic Perspective

Modern age is the age of Science. Science has played the vital role in the making of modern society. Modern science has contributed to the making of modern society by way of developing the technological known how. In the other words, the main thrust of modern science has been towards the *saṁskāra* of material things. The modern technique of refinement, modification, purification and processing the material things is known in Vedic science as *saṁskāra.* Thus we can say that modern science has given stress upon the *saṁskāra* of material things and developed a technology to promote the convenience and comfort of human beings. Here there is a basic difference in the concept of Vedic Science and modern science. Vedic science gave more thrust on the *saṁskāra* of human beings as compare to material things. According to the Vedic science, a human being who has not gone through the certain process of *saṁskāras* remains in his crude form and is recognized as no better than an animal. As such a human being is not considered to be the actual social being. In material things also we can see, so long as they are in their crude form, they cannot be used practically. For their practical use, the material things are to undergo a particular *saṁskāra* or processing. For instance, an oil in its crude form cannot be used, but after undergoing the process of refinement it becomes usable. Thus the conversion of oil from its crude form to the refined form is known as Science. Similarly processing of other material things for the use of human-beings is known as Science. In the race for developing technology modern Science has totally ignored the need for *saṁskāra* of humans. That is why, today in the age of machinery, human beings are also considered to be as machines. Today the human beings are considered as life less things and so there is always a talk about human resource development never as human development. The concept of human resource development lies in considering humans as resources just like other natural resources. Natural resources are often exploited for the benefit of humankind and so are human beings. However, there is basic difference in the exploitation of natural sources and exploitation of human resources. Exploitation of natural resources may benefit the entire humankind, on the other hand exploitation of human resources is not always going to benefit the entire humankind. Exploitation of human resources turns usually in the exploitation of certain helpless human beings for the larger interest of certain opulent group of human-beings.

So far as Vedas are concerned the prime duty of science is not only the *saṁskāra* of material things, but the *saṁskāra* of human beings also. Vedic seers never considered human beings as the means or resources. For them the human beings are not the means but ends. Every thing was directed towards the development and elevation of human beings.

Vedic seers had the humanitarian and ethical approach. Their main aim was to the relieve the human beings permanently of the suffering, deprivation and all other problems that haunted them.

Modern Science has though developed so many aids for the comfort of human beings, yet it has a limited approach. Its approach is not Universal. Modern science has no moral, ethical or humanitarian foundation. It is founded on the laws of survival of fittest in the struggle that ensues among human beings on the globe. Modern science wants to define and describe everything in the light of struggle. It has no remedy to put an end to this struggle. Contrary to this, main aim of Vedic science is to put an end to this struggle and prepare a stage for friendship, fraternity and co-existence of each and every being on the Earth. It clears a stage for co-existence of each and every human-being by giving a call for accepting the principals of mutual understanding and co-operation among human beings.

saha nau bhunaktu saha vīryaṁ karvāvahai

tejasvināvadhitamastu, mā vidviṣāvahai.

"Let there be mutual co-operation in eating and gaining power. Let each one of us become illumined with knowledge. let us not envy or struggle with each other."

Since the modern science has its foundation laid out on the laws of struggle its talk about the exploitation of natural as well as human resources aims at the comfort of those who are able to survive the struggle on the globe. We can see that in spite of high advancement in the area of technology and production of each and everything, the benefits are not reaching to each and every human being on the globe due solely to the absence of moral and humanitarian values. We have more than sufficient amount of food grains piled up in our stocks, still there is poverty and hunger on the globe. At one hand, the excess food grains are being disposed off in the sea, on the other hand human-beings and other animate beings are dying of hunger. The Science has blessed the modern humanity with great prosperity still more than 50% of human beings on the globe are below poverty line and suffering from hunger, thrust, malnutrition, phenomenon of coldness and hotness and dearth of essential things for their survival. One can say unhesitatingly that modern Science has failed in performing the *saṁskāra* of human beings. It could develop and produce things but not real human beings.

Everything has been commercialized. In the race of commercialization moral values have left far behind. Money has become the basis of material advancement but moral values are the real basis of the advancement of humanity on the globe. These moral values are called by *Vedic* seers as *Dharma. Dhāraṇāt Dharma prāhuḥ i.e.* since these values are sustaining human life on the globe, they are called Dharma. Following moral values are said to be the constituents of *dharma.*

Dhṛti (patience), *kṣmā* (forgiveness), *damaḥ* (to have control over one's mind), *asteya* (non stealing), *śauca* (inner and outer cleanliness), *indrīya nigraha* (to have control over one's sensual desires) *dhi* (use of intellect, rationale) *vidyā* (education) *satya* (seeking truth) and *krodha* (calmness, i.e absence of anger). Today we see *Dharma* is applied in altered sense. Different sects or schools propagating and fanning fundamentalism are known as religion, which is not dharma in factual sense.

The saṁśkāra of human beings takes place by inculcating the

above components of *Dharma* (i.e. moral values of life)

There is clear cut observation in the Vedic philosophy that every human being is born as śudra from the womb of his / her mother. After undergoing the process of saṁśkāra, he becomes dvija (twice born).

janmamā jāyate śudraḥ
saṁśkāraiḥ dvija juccyate.

That is a human-being born physically is not better than an animal. The saṁśkāra (inculcation of moral values and education) makes him a real being. For being the actual human being he has to take second birth, i.e. in addition to his physical birth from the mother's womb. He has to take birth form *Ācārya's* womb. *Ācārya* is a person who educate him and help inculcation of moral and humanitarian values.

Acārya kasmāt ācinoti arthān ācāram grāhayatīti sataḥ.

Just like a child stay for nine months in his / her mother's womb, similarly he has to stay with *ācārya* in *Gurukula* for years to be born again as a perfect, knowledgeable social human-being. The stay of child with ācārya is called as *antaḥvāsa* and the child observing *antaḥvāsa* (staying in the womb of *ācārya*) is known as *antaḥvāsī.* Just as in developing a certain instrument or machine, several types of processing is required, similarly in developing a perfect human being, the Vedic sages stressed the need of sixteen *saṁśkāras* right from the birth till death of an individual. Modern Science's advancement is lop-sided. It has played pivotal role towards developing perfect machinery and has cared less in the development of perfect human beings.

So far we have confined ourselves to the physical and mental *saṁśkāra* of the human beings. For maintaining the physical health, researches are going on, but for the upkeep and maintenance of moral health or character, building no research or effort has been put in. Whereas Vedic scientist was equally concerned about the moral (education and character building) as was with maintaining the physical and mental health.

Vedic ideas are still very much relevant for the upkeep and advancement of modern society. The need of hour even today is the moral and spiritual education, inculcation of ethical, moral and humanitarian values and spiritual advancement, so that the modern society may rid of vices of struggle, exploitation, commercialisation of each and every thing and may be endowed with moral / humanitarian and spiritual values, so that all human beings may turn into perfect human beings *(Āryas)* and remove hunger, poverty, depravedness from the globe through their mutual co-operation instead of being compartmentalised and divided along the lines of cast, race, region or religion. Let the whole world sing the Vedic song.

Bhumi mātā putro 'ahaṁ pṛthivyāḥ

"Earth is my mother and I am the child of mother earth."

5

Lost Vedic Scientific Literature

Bharat, the cradle of the world civilisations on the earth, is the place of origin of Vedas and through them the place of origin of all other sciences of the world. It is proved by the fact that from Vedic period till the time of *Mahābhārata* war, science and technology was at its pinnacle. The seers of this land discovered all the laws underlying cosmic and physical creation and thereby developed a vast study material (literature) covering both the aspects of science, i.e. spiritual and material (astrophysical & physical) sciences.

Who affords to forget those dark events of the human history and civilisation, when libraries at Nalanda, Vikramshila and Ujjain were set ablaze? Alexandrian library in west too met the same fate at the hands of the enemies of civilisation. Had all the literature been preserved today intactly, there would have been a different world in terms of history and sciences. Inspite of this mass scale destruction of the books in the libraries, huge number of Sanskrit manuscripts are still preserved in various museums and libraries of India. What to say of India, the libraries and museums of America and Europe are also holding around 1.5 million Sanskrit Manuscripts. The need is to collect, edit and and publish this vast literature.

One more gruelling thing is that there is misconception among learned and laity that Sanskrit is a language of Sanskaras or rituals. Its literature generally deals with spirituality, philosophy, ethics, morality and religion. In 19th centuary, when Swami Dayananda Sarasvati pronounced that Vedas are the storehouses of all true sciences, no body, even the Sanskrit scholars in India were ready to believe this statement. Till 20 years back almost 99% of Sanskrit scholars were not able to telerate that Vedas or Sanskrit contains positive sciences. Until date whatever scientific work is done on Vedas or in Sanskrit language, scholars from the scientific streams have taken the lead. Most of the Sanskrit scholars in India due to their poor background of science, do not dare to take up any scientific studies. Moreover, the Sanskrit scholars in the west do not want to talk about science in Sanskrit out of prejudieces and preconceived notions. They have a prejudiced thinking that the science is the prerogative of the modern west and the ancient language like Sanskrit have nothing to do with science. They associate past with darkness and primitive races. At the most, they take Vedic literature granted for religious or sacred literature. They forget that Sanskrit literature has never propagated religion at any stage or at any time and place. Sanskrit literature talks about Dharma and that Dharma is not different from today's science.

Today when the scholars have started thinking in terms of science in Vedas and other Sanskrit literature, the things have statrted changing. Modern scholars and scientists are taking interest in the scientific studies carried out in the Vedic and post Vedic period. During their researches they have been able to locate a vast body of scientific literature written is Sanskrit during the times of yore. The various scientific studies carried out in Vedas and allied literature and publication of scientific works in Sanskrit has made scholars more curious to search for more and more Vedic scientific literature written in the past. This search can be bi-directional. First thing is to ransack the entire collection of ancient Sanskrit manuscripts located in the libraries and museums of the various countries of the world. Secondly, to locate references of scientific books in the available vast body of Sanskrit literature. The aim of the present paper is to attarct the attention of the scholars interested in the study of Vedic scientific literature to collect maximum possible information in the form of manuscripts from the various libraries and museaums of world and maximum possible references of scientific titles from the extant Sanskrit Texts and their commentaries. Here we present a list of such scientific titles with or without their authors as are referred to in varoius Sanskrit Texts and Commentaries. With the help of this list scholars and readers of Vedic Sciences will be able to have an estimate of the numbers and types of the books written covering the various aspects of science and technology prevalent during Vedic times and thereafter (right from Vedas till the

Mahābhārata).

Agriculture

1. *Agatattvalaharī* -- Deals with Agriculture, methods of cultivation of plant kingdom, description of trees and their treatment. Authorship assigned to Maharishi Atri and Āśvalāyana.

2. *Kṛṣi Parāśara* - Methods of cultivation are discussed here. Authorship is assigned to Parāśara.

3. *Kṛṣi Śāstra* 4. *Kṛṣi Kaumudī*

Plants

1. *Udbhijjatattvasārāyaṇam* 2. *Udbhija Prakaraṇa*

Water

1. *Aptattatva Prakaraṇa* -- Deals with the different types of waters and their utility. Importance of bathing in different waters is mentioned. Characteristic features of different waters is described. Authorship assigned to Maharishi Āśvalāyana.

2. *Aptattva*

Cosmology/Cosmogony

1. *Aṇḍa Kaustubham* - Description of galaxies and types of living beings therein etc. is mentioned. Authorship assigned to Maharishi Parāśara.

2. *Ākāśa Tantra* - Deals with seven types of skies, different portions of universe, clasification of stars in sky, interaction of various energies in sky, types of Power, Fire, Light, orbit of planets, Earth. Rivers and their description is also mentioned. Authorship is assigned to Maharishi Bhāradwāja.

3. *Kaumudī* - The universe is critically discussed in this work. Authorship is assigned to Somanātha.

4. *Kheṭa Sarvasva* - In this work galaxies and planetary motions have been discussed. Authorship is assigned to Maharishi Jaimīni.

5. *Brahmāṇḍa Sāra* - History of Universe is explained here. Authorships is assigned to Vyāsa.

Photography

1. *Aṁśubodhini* - Deals with the photography, planetary motions, the influence of other interfaces on their motions, light, heat sound, telephony, constructions of aeroplanes, electricity and its applications.

Dictionary of Technical terms

1. *Paribhā|endash ā Candrikā*
2. *Nāmārtha kalpa-sūtram* (Science of naming the different parts of machine): In this 84,00,000 śaktis (energies) have been mentioned with their names. Also as how they can be generated and meaning of words etc, are also described. The authorship is assigned to Maharishi Atri.

Accoustics

1. *Sarvaśabda Nibandhanam* (Acoustics)

Ayurveda

1. *Ṛk hṛdaya Tantra* - Diferent types of diseases, their treatments is described therein. Authorship assigned to Maharishi Atri.
2. *Ayurveda Prakāśa*

Arthaveda

Arthaveda

Animal Sciences

1. *Aśva Śāstra* 2. *Gaja Śāstra* 3. *Aśva Lakṣaṇa Sāra*
4. *Mṛga cāramiya* 5. *Hayalīlāvati* 6. *Go Śāstra*
7. *Kukkuṭa Śāstra* 8. *Aśva Tantra* 9. *Gaja Tantra*
10. *Gorakṣā Tantra*

Dhanurveda

1. *Dhanurveda by Maharishi Vasiṣṭha*
2. *Astra Vidyā* 3. *Vyuha Śāstra* 4. *Senā Śāstra*
5. *Rathaśīkṣāśāstra* 6. *Sūta Śāstra*
7. *Vāhanarohaṇa Śāstra*

8. *Kamandaka* 9. *Dhanu Śāstra* 10. *Bhojarājīva*

11. *Revantottara* 12. *Vyuhua Lakṣaṇa* 13. *Senā Lakṣaṇa*

Marshal Arts

1. *Malla Śāstra* 2. *Malla Vidyā Prakāśa*

3. *Weapons/Armoury* 4. *Nālikānirṇaya*

Space

1. *Kheṭa-vilasa-grantha (Flying in sky)*
2. *Ākāśatantram of Bharadwāja*

Fire

1. *Vaiśvānara Tantra* - 128 types of fire, their colours, behaviour, uses, measurements, mutual differences etc. are described. Authorship is assigned to Maharishi Nārada.

Mechanics/Machinery

1. *Śuddha Vidyākalpa* - This describes different types of machines including for making an aeroplane and sound machines. The works done by siron, war with Robot-dolls. Dances, music and gate keeping by war dolls. Authorship is assingned to Āśvalāyana.
2. *Bṛhadyantra sarvasva* - Explains all types of machines including that of aeroplanes and electricity. Authorship is assingned to Maharishi Bharadwāja.
3. *Yantra Kalpa* (Flying Machines) by *Maharishi Garga*
4. *Yantra Saṅgraha*
5. *Yantra Kalpataru* (Mechanical Engineering)

Aeronautics

1. *Samarāṅgaṇa Sūtradhāra* - This describes the manufacturing of airplanes with mercury as their fuel. Authorship is assgined to King Bhoja of Malwa.
2. *Vimāna Candrikā by Nārāyaṇa*
3. *Vyomayāna Tantra by Maharishi Śaunaka*
4. *Yāna bindu by Vacaspati*

5. *Kheṭayāna Pradīpikā}* by *Cakrāyaṇa*
6. *Vyoma-yāna-arka-prakāśa* by *Ḍhuṇḍīnātha*
7. *Vyoma-yānārka Sikṣā}*
8. *Kriyāsāra (Practical exercises to the Aeronautics)*
9. *Śaunakīyam (Work of Śaunaka on airplanes)*
10. *Kheṭa sarvasva* *11. Kheṭa Yantram*

Instrumentation

1. *Yantraprakaraṇa (Installation of instruments in planes)*
2. *Maṇiratnākara*
3. *Maṇibhadra Kārikā} (Installation of Maṇis in Planes)*
4. *Maṇi Prakaraṇa (Installation of Maṇis in Planes)*
5. *Maṇi Kalpa Pradīpikā}*

Airspying

1. *Cāra-Nibandana grantha*

Airconditioning in Aeroplanes

1. *Ṛtukalpa*

Colour

1. *Varṇsarvasvam* (Colour-management)

Meteorology

1. *Karaka Prakaraṇa* - Deals with changes in clouds, changes in sun-rays, and relationship of clouds. It also discusses the role of sun-rays in generating precious stones. (Navaratnas). Authorship is assigned to Aṅgirasa.
2. *Meghotpatti Prakarṇa* - Types of clouds, thunder, lightning, and their effects are explained here. The changes in clouds, generation of the life of many species, changes in solar energy, relationship between solar radiation and the cloud formation, orign of navaratnas and how the solar radiation is responsible for their origin. Authorship is assigned to Maharishi Aṅgirassa.
3. *Saudāminī Kalā* (Science of lightning)

Metallurgy

1. *DhOātu Sarvasva* - In this work elements, minerals, metals, alloys and their extraction from mines by different methods are discussed. This also deals with poisons and antidots, production of mercury (since mercury was also used as a fuel in machinery), shulphur etc and preparations of ashes. Authorship is assigned to Maharishi Baudhāyana.

2. *Lauha Tantra* - Ores, their genesis and extraction are discussed here. Authorship is assigned to Śākaṭāyana.

3. *Lauha Tattva Prakaraṇa*4. *Lauha Ratnākara*

5. *Lauha Rahasyam* 6. *Lauha Śāstram of Śākaṭāyana*

7. *Lauha Paddhati*

8. *Lohasarsvam (Work on metallurgy of planes)*

9. *Dhātusarvasvam (Work on aeronautical metallurgy)*

11. *Lohaprakaraṇa (Work on metallurgy of planes)*

12. *Lauhādhikaraṇam*

Science of Smoke/Vapour

1. *Dhuma Prakaraṇa* - This work discusses in detail different types of smokes/vapours. The detection of vapours/smokes with the help of mirrors. Research on various types of vapours/smokes with acids to find out whether the vapour is harmful or not for mental and physical growth. Authorship is assigned to Nārada.

2. *Āpa Tattva* - In this 84,000 vapours, their layers, their impact on earth and plantation, 84,00,000 medicines and instruments to detect these vapurs are described. The authorship is assigned to Maharṣi Śākṭāyana.

Creation

1. *Prapañca Laharī* - This work discusses atom in detail. The question whether this Universe is created by atoms or by Brahma tattva is addressed here. Authorship is assigned to Vasiṣṭha.

2. *Sṛṣṭi Vilāsa*

3. *Prapañcasāra*

Language

Loka Saṅgraha - This work discusses 1714 languages, living beings, their origin, and food habits and different informtions of the world. Authorship is assigned to Vivarṇācārya.

Air

Yāyu Tattva Prakaraṇa - This work discusses ionosphere, different layers in it. 4000 varieties of gases and effects of these gases on the earth, on flora and fauna. The instruments to detect them in atmosphere are explained in detail. Authorship is assigned to Maharishi Śākaṭāyana.

Chemistry

1. *Rasa-ratna-samuccaya* - This work discusses chemistry. Different chemicals like mixed chemicals, compound chemicals, ordinary chemicals, their actions and reactions with different metals have been discussed. Different machineries, metallurgy, elements and their characters have been explained. Metals like Gold and Silver etc. their characters and their different kind of treatments to humanbody has been discussed in detail. Authorship assigned to Vāgabhaṭṭācārya.
2. *Rasa Ratnākara-* This books is divided into five *Khaṇḍas :* 1. *Rasa Khaṇḍa* 2. *Rasāyaṇa Khaṇḍa* 3. *Rasendra Khaṇḍa* 4. *Vāda Khaṇḍa* 5. *Mantra Khaṇḍa.* The authorship is assigned to *Nityanātha Siddha.*
3. *Rasendra Cuḍāmaṇī by Somadeva.* 4. *Rasendra Cintāmaṇi*
5. *Rasa Saṅketa Kārikā|tab*
6. *Rasendra Maṅgalam by Nagarhjuna*
7. *Rasa Sāra* 8. *Rasanāmadhenu*
9. *Rasa Kaumudi* 10. *Rasendra Vijñāna*
11. *Rasendra Sāra Saṁgraha* 12. *Rasādhyāya*
13. *Rasopaniṣat* 14. *Ānanda-kāṇḍa*
15. *Rasa-paddhati* 16. *Rasa Kāmadhenu*

Power/Electricity

1. *Śakti Tantra* - This describes electricity and its powers such as

sarvākarṣa, rūpākarṣa, rasākarṣa, gandhākarṣa, śabdākarṣa, dhairyākarṣa, śarīrākarṣa, prāṇākarṣa, and others (a total of 16 types of powers) are explained. Authorship is assigned to Maharishi Agastya.

2. *Nāmārdha Kalpa* - This work defines 8.4 Million types of powers and gives their nomenclatures. Authorship is assigned to Maharishi Atri.
3. *Śakti Bījam* (Power Supply in Machines)
4. *Śakti Kaustubha* (Power Management)
5. *Śakti Sarvasva* (Power management in Planes)
6. *Śakti Vilāsa*
7. *Śakti tantram (Work on Power management in planes)*
8. *Śaktisūtram of Agastya*
9. *Rūpaśakti Prakaraṇam of Aṅgiras*

Speed Management in Machines

1. *Saṁskāra Ratnākara (Speed management in Aeroplanes)*
2. *Gatinirṇayādhikāra*
3. *Drāvaka Prakaraṇa*
4. *Gatinirṇayādhyāya (Speed management)*

Sound Management in Machines

1. *Śabda Mahodadhi (Sound management in Aeroplanes)*
2. *Śabda Nibandhanam (Sound controle in machines)*

Science of Rays

1. *Āśani Kalpa*
2. *Aṁśum Tantram of Bharadwāja*
3. *Aṁśubodhinī*

Glass Technology

1. *Darpaṇa Kalpa*
2. *Darpaṇa Śāstram*
3. *Darpaṇa Prakaraṇa (Mirror Planning in Aeroplanes)*

4. *Mukura Kalpa (Mirror Technology)*
5. *Darpaṇa Saṅgraha (Mirrology)*

Witchcraft

Sammoha Kriyākāṇḍam (How to make an enemy unconscious)

Gemology

1. *Suvarṇa Ratnādi Parikṣā\tab* 2. *Navaratna Lakṣaṇa*

Commerce

Vāṇijya Śāstra

Geology

Bhū Parikṣā Śāstra

Ethics

1. *Nīti Śāstra* 2. *Nīti Sāra* 3. *Nīti Sūtra*
4. *Nīti Kalpataru* 5. *Nīti Prakāśikā\tab* 6. *Nīti Mañjarī*
7. *Nīti Candrikā*

Military Science

Samara Sāra

Pharmacy

Auṣadhi Kalpa of Atri

Artha Śāstra

1. *Artha Śāstra by Kauṭilya*
2. *Artha Śāstra by Bṛhaspati*
3. *Artha Śāstra by Bharatṛhari*
4. *Rājanīti Ratnākara*

Political Science

1. *Nāgara Sarvasva* 2. *Daṇḍa Nīti Śāstra*
3. *Dhaumaya Nīti Śāstra*
4. *Akṣnīti Sudhākara Gaṇita (Mathematics)*

1. *Aṅkagaṇita* 2. *Bījagaṇita*
3. *Kṣetragaṇita* 4. *Chāyāgaṇita*
5. *Kūpādigaṇita* *6. Bhinnādhikāra*
7. *Prakīrṇagaṇita* *8. Trairasikagaṇita*
9. *Miṣragaṇita* 10. *Ghaṭagaṇita*
10. *Vāstugaṇita* 12. *Sūtragaṇita*
13. *Suvarṇagaṇita*
14. *Vālamīki gaṇita*

Architecture

1. *Brahmāṇḍa Vāstu*
2. *Rājadhāni Nagar Nirmāṇa vāstu*
3. Pattana Vāstu *3. Grāma vāstu*
4. Nāvika Vāstu *5. Sthāpatya Tantra*

Geography

1. *Bhūgola* 2. *Dvīpa Viveka*

Taxicology

Viṣanirṇayādhikāra (Air -Texicology)

Food Technology

Aśana kalpa (Food technology)

Tempering Technology

Pāka-sarvasva *2. Pārthiva-pāka-kalpa*

Saṁskāra Darpaṇa (Tempering and heating technology)

Thermodynamics

Bhastrikānibandhanam

Textile Engineering

Paṭakalpa *3. Kṣirīpaṭakalpa*

Paṭapradīpikā\tab *4. Paṭṭasaṁskāra Ratnākara*

(Textile technology)

Miscellaneous Books

1. *Drāvaka Prakaraṇa* 2.
3. *Śucivaṇa Karma* 4. *Nidhi Pradīpa*
5. *Abhilaṣitārtha Cintāmaṇī* 6. *Yukti kalpataru*
7. *Paribhā\endash ā Candrikā\tab*
8. *Nāmartu Kalpa by Atri*
9. *Dhvānta Vijñāna Bhāskara*
10. *Akṣa Tantra by Atri*
11. *Nāgārjuna Tantra by Aiśvara*
12. *Māṇḍvya Tantra* 13. *Vyāḍi Tantra*
14. *Patañjali Tantra* 15. *Kapiñjala Tantra*
16. *Bhṛgu Tantra* 17. *Agastya Tantra*
18. *Aprājita Pṛcchā\tab*
19. *Aneka Vidyākalpa Nirupaṇādhyāya*
20. *Arya Vidyā Sudhākara* 21. *Mānasāra*
22. *Mānasollāsa* 23. *Sāmarājya Lakṣmī Pī\emdash aka*
24. *Avyāja Palini Tantra* 25. *Rajyalakṣamī prasādaka Tantra*
26. *Aṅgarāgamana Tantra (How to walk on the burning coals)*
27. *Rājyopāṅga tantra* 28. *Stribalarāja Tantra*
29. *Cāmaradvandva Tantra* 30. *Praharṣarāja Tantra*
31. *Dhīra Tantra* 32. *Grāmapālana Tantra*
33. *Kāmarūpa Tantra* 34. *Kālajivanika Tantra*
35. *Kuñja Kambala Tantra* 36. *Kanakaprayoga Tantra*
37. *Bodhini Tantra* 38. *Pādukā Tantra*
39. *Kheṭaka Tantra* 40. *Lepa Tantra*
41. *Śalya Tantra* 42. *Pātāla Gamana Tantra*

43. Cakrāyuddha Prayoga Tantra

44. Lekhā Tantra *45. Pṛthivi Tantra*

46. Śulka Tantra *47. Giri Tantra*

48. Vana Tantra

50. Bodhānanda Kārikā of Bodhānanda

51. Viśvambhara Kārikā of Viśvambhara

52. Mūlārkaprakāśikā\tab *53. Śaṇaniryāsa Candrikā*

53. Bṛhatkāṇḍam *55. Paṭikānibandhana*

56. Nirṇyādhikāra (Decision making)

57. Mū\endash a Kalpa *58. Kuṇḍakalpa*

59. Kuṇḍanirṇaya *60. Parāṅkuśa*

61. Niryāsa-kalpa *62. Pralaya Paṭalam*

63. Ṣaḍagarbha viveka *64. Raghūdaya*

65. Jīva sarvasvam of Jaimini

66. Karmābdhipāra of Āpastamba

67. Chandaḥ Kaustubha of Parāśara

68. Kaumudī of Siṁhakhoṭha

69. Ṛkhṛdayam of Atri

70. Lokasaṅgraha of Visaraṇa

Apart from the abovecited treasure of Vedic scientific literature, we also come across the names of some 36 eminent Vedic scientists who contributed brilliantly in various scientific fields during the Vedic period. They may be enumerated as under:

1. Nārāyaṇa Muni	2. Śaunaka	3. Garga
4. Vécaspati	5. Chākrāyaṇi	6. Dhuṇḍinātha
7. Viśvanātha	8. Gautama	9. Lalla
10. Viśvambhara	11. Agstya	12. Buḍila

13. Gobhila	14. Śākaṭāyana	15. Atri
14. Kapardī	17. Gālava	18. Agnimitra
15. Vātāpa	20. Sémba	21. Bodhānanda
22. Bhardwāja	23. Sidhnātha	24. Īśvara
25. Āśvalāyana	26. Vyāsa	27. Parāśara
26. Siṁhakoṭa	29. Aṅgirā\tab	30. Visaraṇa
27. Vasiṣtha	32. Jaimini	33. Āpastamba
34. Baudhāyana	35. Nārada	36. Vālmīki

www.ingramcontent.com/pod-product-compliance
Lightning Source LLC
Chambersburg PA
CBHW032021190326
41520CB00007B/567
9788187710639